MUSINGS OF THE CONTEMPLATIVE LEADER

A leadership manifesto for strategic and digital leaders in this digital era

Satya Sanivarapu

SANSID BOOKS
Seeing Clearly

Published by SANSID BOOKS

Seeing Clearly

ISBN: 979-8-9943296-9-6 (paperback)

ISBN: 979-8-9943296-8-9 (ebook)

First edition

Printed in the United States of America

Preface

There is largescale 'Digital Disintermediation' rampant in leadership circles today. Digital's evolution has outpaced its comprehension and is the elephant in today's boardrooms. There is pressure to harness Digital on scale for exponential leverage in every industry and organization.

However, the context on Digital is dependent on education, foundation, aptitude, and experience that forms the lens of its perception.
How Digital 'sits' in the minds of Leaders is tightly coupled to the extent of its harnessing. A comprehension of the disintermediation is imperative.

This pocketbook aims to stir the foundations and put Digital in context for the 'Contemplative Leader'.

It is framed in Six short chapters, packed with a path that fully assimilates in thought and action with time. This is a book to consume, pocket, and ponder.

The audience for this book is anyone who wears the Leadership hat in this modern age - one with outcomes to achieve with the most important enabler - Digital. This book is for each of you.

The recommended way to read is Slowly, Reflectively, Recursively. The destination of Digital Transformation is the Quiet, Confident, Empowerment of People, Teams, and Partners in the Value Chain.

Table of Contents

CHAPTER ONE

DIGITAL: THE ELEPHANT IN THE BOARDROOM

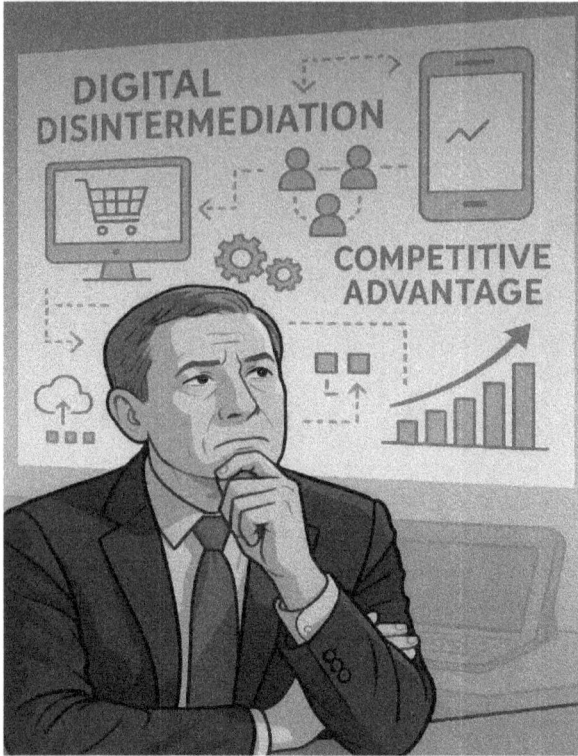

In today's crowded digital marketplace—where every tool, service, and solution is branded as a game-changer—how should the 'Contemplative Leader' make sense of it all? How do we cut through the noise, make sound decisions, and still shield our organizations from the relentless competitive churn?

This book opens a reflective dialogue to tackle the most loaded question in modern enterprise: *What does Digital really mean?*

Digital: More Than a Tool?

For many, the word "Digital" conjures up technology, tools, and transformation buzzwords. But leadership perceptions are often shaped by years of operational experience, fragmented knowledge, and market conditioning. In most boardrooms, digital is still treated like a Swiss Army knife—useful, but just another option on the table.

The truth? Digital isn't just a tool. It's a paradigm. And leaders can no longer afford to treat it as a plug-and-play solution for isolated problems. The pace of change—and the semantic overload it brings—has blurred the very definition of what digital entails. This makes clarity not just helpful, but essential.

Reframing Digital: A Contemplative Lens

For the Contemplative Leader, digital is not a noun, but a framework. One that is understood best through two key dimensions:

1. The Value Timescale:

Here, digital investments are mapped against short-term and long-term value.

- Tactical wins include automating a routine process, creating dashboards for operational visibility, or setting up anomaly alerts in a supply chain.
- Strategic wins aim higher—building agile supply chains, launching new channels, or enabling entire business model shifts.

2. Resource Enablement:

Digital shines when it enhances the capabilities of core resources—people, processes, and assets.

- People thrive when collaboration becomes seamless—across teams, customers, and partners.
- Processes evolve through automation and smart exception handling.
- Assets deliver greater **Return on Investment (ROI)** when guided by predictive planning and intelligent maintenance.

When viewed this way, digital becomes more than a cost center or buzzword—it becomes a multiplier.

The Leadership Imperative

Triggering this inner dialogue helps leaders align digital decisions with broader organizational purpose. The Contemplative Leader knows that:

- Small wins matter—but only if they stack up to long-term gains.
- Empowered resources create resilient systems.
- Strategic clarity around digital leads to smarter, faster decision-making.

This book will continue to peel back layers of digital ambiguity, offering pragmatic lenses to help leaders think critically, act confidently, and lead wisely in the age of digital acceleration.

CHAPTER TWO

DIGITAL TRANSFORMATION: FIRST STEPS FROM THE TOP

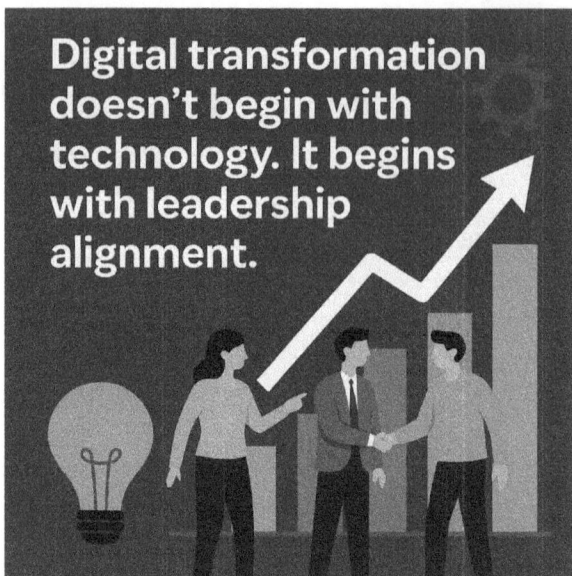

Digital transformation doesn't begin with technology. It begins with leadership alignment.

In Chapter One, we reframed Digital through the twin-lenses of **Value Timescale** and **Resource Enablement**.

With this foundation in place, the Contemplative Leader now looks inward—toward the circle of immediate influence—to chart a path forward.

Digital transformation doesn't begin with technology. It begins with leadership alignment. The first step? Engage your leadership team in deliberate dialogue around governance, policies, standards, and **Key Performance Indicators** (KPIs)—all framed in the context of data and digital tools. These form the backbone of a transformation that's not just flashy, but functional.

Start with the End in Mind

True digital transformation must serve the organization's strategic vision—not distract from it. Whether your ambition is to become the distributor of choice, the market leader in sustainability, or the go-to for customer experience excellence, your digital roadmap must reflect your business context—both macro and micro. That means defining the capability bridge between where you are and where you need to be:

- **Want to lead across regions?** Invest in advanced inventory management and integrated planning platforms to match demand by market.

- **Aiming to deepen customer loyalty?** Build out a cloud-native **Customer Relationship Management** (CRM – like Salesforce, Microsoft Dynamics 365) powered by **Artificial Intelligence** (AI), tightly integrated with your **Enterprise Resource Planning** (ERP – like SAP S/4 Hana, Oracle NetSuite) and **e-commerce** stack of applications.

- **Delivering faster product turnaround for your strategic items?** That may mean reimagining design processes for supply chain agility and forming digital partnerships

with your strategic suppliers
sharing information in real-time.

Sometimes, the most powerful digital moves are non-obvious. But always, they are tied directly to the organization's unique goals and constraints.

Governance: The Anchor of Accountability

Once vision and capabilities are aligned, the next focus is governance—the often-overlooked layer that determines whether your digital strategy scales or stalls. Here's how the Contemplative Leader sets the tone:

- **Break the silos.** Propose data-sharing policies that clarify ownership across systems and functions. Who owns the data? Who manages the tech? Who's accountable for customer outcomes?

- **Reframe Key Performance Indicators (KPIs).** Revisit metrics to ensure they support true business value. A warehouse with 90% utilization may sound efficient—until it becomes a bottleneck during demand surges. A fast lead time may help...unless it inflates logistics costs due to over-frequent deliveries.

- **Standardize data language.** Define how departments share and interpret data—down to units of measure, conversion rules, and time-based context. Mismatched definitions can derail even the best intentions.

- **Enforce data quality policies.** Good data is not a happy accident. It's an outcome of structured policy and proactive hygiene. Embed audits, alerts, and real-time validations. Catch the errors before the damage is done—be it overbooked trucks, invoice mismatches, or incorrect units of measure.

A Thoughtful Start Is a Strategic Start

The early steps of digital transformation are less about bold moves and more about quiet clarity—the kind that prevents chaos later. By **establishing governance, aligning outcomes to vision, and demystifying data flow,** the Contemplative Leader lays a resilient foundation.

Digital transformation is not a sprint to software adoption. It is an organizational awakening—rooted in intentionality, built on insight, and led from the top.

CHAPTER THREE

FROM INTENTION TO IMPACT: DRIVING MOMENTUM IN DIGITAL INITIATIVES

Digital transformation isn't just about strategy—it's about strategic movement.

We previously defined digital in context—viewing it through the lens of **Value Timescale** and **Resource Enablement**—and explored how leaders can ground transformation through governance and policy.

Now, we move from principle to practice. What does it take to turn digital vision into meaningful action?

The Contemplative Leader begins by taking stock—then daring to reimagine.

1. Evaluate the "As-Is"

Every transformation begins with self-awareness. Map the current state—not just processes, but the technology architecture enabling them.

Ask:

- Is the **Customer Relationship Management** (CRM) system integrated with your **Enterprise Resource, Planning** (ERP) system?

- Can demand signals be traced, translated, and trusted across the supply chain?

- Do exceptions trigger real-time alerts for the right teams—or get buried in silos?

These questions illuminate gaps and reveal whether the systems and processes are wired for a Connected Business. Without this foundation, transformation becomes aspiration without traction.

2. Define the "To-Be"

Next, visualize the ideal future—without being tethered to current limitations.

- What capabilities must there be?

- What experiences are to enable?

- Where does cloud infrastructure (public or private) fit into the puzzle?

At this stage, don't let today's tech constraints limit tomorrow's possibilities. The real barrier is rarely the availability of technology—it's the ability to harness it. That gap is the digital investment—and your journey.

3. Make Change Manageable

Most leaders acknowledge that change management is critical—but few have a consistent playbook. Every transformation is different, but certain principles can become catalysts.

Here's how the Contemplative Leader mobilizes momentum:

Council of Digital Champions

Form a core team of internal "Digital Champions"—individuals who understand how technology enables business processes. This council becomes the nucleus of change, helping seed transformation across departments and functions.

Their charge isn't to implement tech—it's to transform outcomes through tech. Their perspective is grounded in process and powered by technology.

Alignment Across People, Process, and Assets

Alignment is the beating heart of transformation.

- **People:** Equip teams with clarity on standards, policies, and vision. Educate through micro-actions— clear metrics, collaborative workshops, and inclusive tech onboarding. Start with the skeptics. Win them over. Resistance, when engaged, becomes momentum.

- **Processes:** Fuse human and machine intelligence. Shift from micromanagement to manage-by-exception. Design processes that flag what matters and empower teams to act decisively.

- **Assets:** Whether software or physical infrastructure, every asset must serve people and process.

Software is underutilized in most firms—its real ROI is unlocked only when functionality is configured, adopted, and measured. Likewise, physical assets drive transformation when repurposed to eliminate failure points, reduce cycle time, and drive smarter decisions.

From Thoughtful Planning to Thoughtful Action

Digital transformation isn't just about strategy—it's about strategic movement. The Contemplative Leader knows that momentum builds from alignment, clarity, and action grounded in business reality.

By evaluating today, defining tomorrow, and stewarding change through champions and principles, transformation becomes more than possible—it becomes inevitable.

CHAPTER FOUR

DEFINING THE YARD STICK: MEASURING PROGRESS OF DIGITAL INITIATIVES

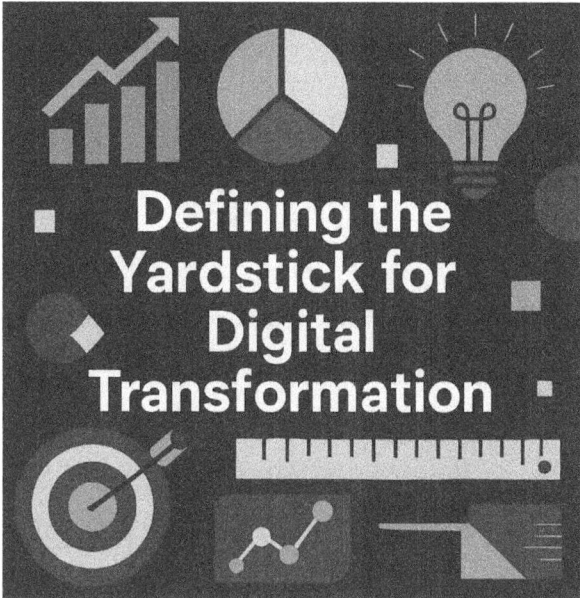

Defining the Yardstick for Digital Transformation

The journey began by contextualizing Digital and its strategic relevance. We then explored how to Organize around digital through Governance and Policy, followed by practical steps to ignite transformation across the enterprise.

Now in Chapter four, we turn focus to something fundamental: how to measure Progress? In other words—what's the yardstick for digital transformation?

Why Measurement Matters

In science, measurement is the baseline for understanding growth, movement, and change. In business, the principle is no different.

If digital transformation is to be more than a buzzword—if it's to deliver sustained prosperity—then leaders must define a clear, actionable way to track its maturity.

This is where the **Digital Maturity Index (DMI)** comes in—a composite metric developed at the organizational level. Think of it as the pulse of your digital journey, aggregating progress across key capabilities that enable business performance.

Designing the Digital Maturity Index (DMI): A Strategic Framework

For the Contemplative Leader, this isn't just a dashboard exercise—it's a planning tool. Next are the foundational components to consider:

Automation & Integration

Track how seamlessly your functions are digitized and connected. Key indicators may include:

- **System Inventory Accuracy:** The percentage match between system data and physical stock. Higher congruence = better integration.

- **Automated Replenishment Rate:** How many restocks are triggered without human input?

- **"No-Touch" Order Percentage:** Orders processed entirely digitally,

segmented by market, channel, or
type.

Digital Customer Engagement

Digital maturity isn't just internal—it's also
measured in how customers engage and
self-serve.

- **Digital Customer Interaction Index:** Captures interaction volumes, response times, and resolution cycles across digital channels (e.g., chatbots, portals, email).

- **Content Engagement Rate:** How users engage with your digital content—alerts, dashboards, or self-service views.

- **Repeat Digital Interaction Rate:** The frequency with which customers return to your digital touchpoints.

Data Analytics

Assess your analytics stack across these dimensions:

- **Descriptive**: Basic visibility and reporting.

- **Predictive**: Forward-looking models and forecasts.

- **Prescriptive**: Optimized recommendations and automated decision support.

Your maturity rises as you move up this hierarchy.

Internet of Things (IoT – physical sensors in the supply chain that are connected to the internet and can send or receive data autonomously**) & Smart Technologies** (systems that can sense, think, decide and act – autonomous systems)

27

Capture adoption and utilization of intelligent tech:

- Real-time tracking sensors

- Smart warehousing systems

- Predictive maintenance alerts for assets and infrastructure

Digitization of Business Processes

Finally, measure the percentage of the end-to-end business processes that are digitally enabled—either via integration platforms, automation, or workflow orchestration tools.

This is the high-level view: how much of your operation runs on digital rails? How much of it is traceable, transparent, and transformative?

Creating a Digital Maturity Index provides more than just visibility — it offers traceability. It helps leadership:

- Justify investments

- Align resources

- Benchmark performance

- Sustain digital transformation initiatives over time

For the Contemplative Leader, it becomes a compass—guiding not just the what, but the how and when of digital progress.

CHAPTER FIVE

UNPACKING THE DIGITAL MATURITY INDEX

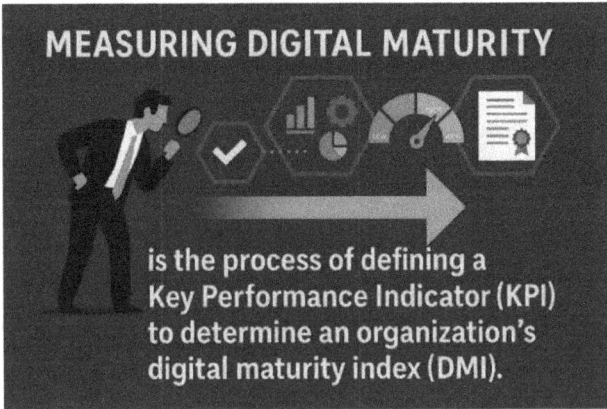

MEASURING DIGITAL MATURITY

is the process of defining a Key Performance Indicator (KPI) to determine an organization's digital maturity index (DMI).

By now, we've gotten knee-deep into the Digital Transformation voyage. We've framed Digital in context, laid the groundwork with Governance and Policy, ignited momentum across the enterprise, and developed a yardstick to measure Digital Progress.

Next step: Building the Digital Yardstick.

This chapter focuses on developing a practical approach to computing the **Digital Maturity Index (DMI)—a Key Performance Indicator (KPI)** that quantifies how far a department, function, or enterprise has progressed on its digital transformation journey. The DMI is an aggregate measure of a basket of KPIs with weights suitably assigned in alignment with strategic objectives. The weights indicate how important of a KPI it is to digital maturity and thus influences the DMI score more heavily.

But before we crunch the numbers, we must first un-learn and re-learn—embracing new ways, looking at digital health through structured and meaningful metrics.

Constructing a sample DMI Component

Let's bring this to life through a tangible example:
System Inventory Accuracy (SIA)—a KPI that reflects the synchronization between what the systems say there is in stock and what physically sits in the warehouse.

Measure Definition: At its core, SIA reveals how well-integrated and aligned your inventory systems are with physical reality. It's both a litmus test for digital maturity and a barometer for operational efficiency.

- Descriptive Analytics: Tracks real-time alignment between system and physical inventory.

- Predictive Analytics: Assesses the probability of mismatch occurring in the future.

Common Causes of Discrepancy

- System: Integration gaps between ERP and inventory system modules

- People/Process: Manual data entry errors

- Partners: Third-party logistics delays

- People/Process: Inaccurate returns or damaged goods restocking

The Business Impact

Discrepancies in inventory accuracy can ripple through the supply chain, creating:

- Misinformation on product availability

- Unpredictable service levels

- Increased planning errors

- Customer dissatisfaction

- Erosion of trust across partners

In contrast, accurate, synchronized inventory data supports agile planning, seamless collaboration, and improved service metrics.

Representing the Measure

- **Method 1**: SKU-Level Accuracy [Accuracy = (Smaller of System Count or Physical Count) ÷ (Larger of the Two)]
  Overall accuracy (SIA) is the average across all SKUs.

- **Method 2**: Aggregate Accuracy [**Overall Accuracy (SIA)** = Total Correct System Counts ÷ Total Physical Count]

- **Typical Range**: Expressed as a percentage (or on a 0 to 1 scale). In practice, 95–98% accuracy is a

strong benchmark in most
industries.

Building Blocks of the Digital Maturity Index (DMI)

System Inventory Accuracy (SIA) is just
one of many DMI building blocks.
Others may include:

- *Automated Replenishment Rate*

- *No-Touch Order Percentage*

- *Digital Customer Interaction Index*

- *Data Analytics Penetration
  (Descriptive → Prescriptive)*

The key? These metrics must be relevant
to your business context. The DMI is not
a one-size-fits-all scorecard.

Context Is Everything

When creating the DMI, consider:

- **Scope:** Will it apply to a single
  function or span the enterprise?

- **Weightage:** Are you assigning
  priority to the right KPIs?

- **Comparability:** Use it to compare progress within your organization over time—not against others.

The DMI is not about benchmarking competitors. It's about holding up a mirror—measuring one's own digital evolution with consistency, honesty, and intent.

Closing Thought

Digital maturity isn't a destination—it's a continuum.
The Digital Maturity Index gives the organization a compass, not just a score. It tells you where one is, where one is headed, and the velocity of getting there.

In the final Chapter, we reflect on how to make the DMI a living, breathing tool—integrated into strategy, not separate from it.

CHAPTER SIX

ROLLING FORWARD: SHAPING MINDSETS OF PEOPLE

Digital transformation should have guardrails—empowerering users to move forward fearlessly.

In the previous chapter, we gave structure to transformation—placing measurement and accountability at the heart of digital progress. Now, we conclude by addressing a subtler, yet equally critical layer: how we think about digital transformation.

Because transformation doesn't just happen through systems—it flows through people.

In this final reflection, we explore the emotional and cultural dimensions of digital adoption. What does it take for employees to trust and embrace change? How do leaders empower teams to navigate risk, build confidence, and use technology fearlessly?

Positioning Digital to Employees: From Mystery to Mastery

To many employees, digital tools are still a mystery. Interfaces evolve rapidly, cryptic error messages abound, and beneath every clean **User-Interface** (UI – the window through which humans command the digital world) lies a perceived jungle of scripts, codes, and potential pitfalls.

This sense of mystery can trigger various fears—fear of clicking the wrong thing, fear of breaking a process, or worse, fear of becoming the weak link in a cyberattack. These fears aren't unfounded and disastrous incidents occur with increasing frequencies as malicious actors leverage weak points in technology.

Enterprises have crippled with various types of attacks.

Ransomware attacks spread through shared systems of access when an **Information Technology** (IT) system (software, shared drive, file server, middleware) of a supply chain partner is compromised.

Data and Integration attacks spread silently corrupting planning, demand, pricing, or availability data when **Application Programming Interfaces**

(APIs- the method for digital systems to communicate in real time), data pipelines, or **Electronic Data Interchange** (EDI – machine to machine exchange of structured business transactions and documents between companies) feeds are attacked. This is very dangerous because it may not trigger any alerts.

Software attacks take place when vendor software is compromised and credentials are stolen, and updates, libraries, APIs are compromised.

These are the realities of today's interconnected digital supply chain ecosystems (systems, data exchange gateways, file servers, IoT sensors) — where security of every node matter, and every user is a gatekeeper.

Safety as a Foundation for Adoption

It's not enough to ask employees to adopt new tools—we must equip them to do so safely and confidently.

This starts with two things:

- Guardrails – **Information Technology** (IT) departments must establish clear protections across the digital supply chain, including vendor and partner interactions.

- Education – Users must be trained not just in tool usage, but in cyber-awareness—spotting phishing emails, verifying secure sites, and treating every digital interaction with care.

When users understand both the protections in place and their role in maintaining them, fear gives way to confidence.

Rethinking Digital: From Intimidation to Empowerment

Digital resistance often stems from fear — of triggering a failure or getting blamed for a system disruption. But today's platforms are built differently.

Modern software is designed with the user at the center. Intelligent workflows, embedded validations, and intuitive interfaces reduce error and build trust.

Gone are the days when a single **Structured Query Language** (SQL - standard computer language written on a computer and used to store, retrieve, analyze, and control data in stored databases) query could crash the ERP. In today's systems, user governance is embedded by design.

When users realize their actions have built-in safeguards—and that mistakes are part of learning—they're more likely to experiment, adopt, and ultimately, champion transformation.

Building Change from the Inside Out

True transformation takes root when users feel ownership.

Let them test the tools. Let them try to "break" them. Involve them early, often, and openly. This is the heart of sustainable change management—when users no longer fear technology, but understand it, trust it, and shape it.

This mindset shift—from fear to curiosity, from intimidation to inclusion—is what truly sets the change management wheels in motion.

A Final Reflection

Digital transformation isn't just about systems.
It's about mindsets.
It's about trust.
It's about putting humans at the heart of the digital journey.

To the Contemplative Leader, the destination was never just digitization—it was the quiet, confident empowerment of people, teams, and partners across the value chain.

Transformation, after all, is not a one-time event.
It's a continuum.

Epilogue

Digital will continue to evolve—faster than governance, faster than comfort, faster than certainty. New abstractions will emerge, new intermediations will dissolve, and new paradigms will quietly take their place. This is not a phase. It is the environment within which modern leadership now operates.

What ultimately distinguishes leaders in this environment is not their proximity to technology, but their posture toward it.

Digital, when approached without reflection, amplifies confusion. When pursued without context, it fragments organizations.

When adopted without restraint, it replaces judgment with motion.

Yet when understood clearly—when situated within purpose, values, and human systems—it becomes a powerful, quiet enabler.

The contemplative leader does not chase Digital.
Nor do they resist it. They place it—deliberately—within the architecture of people, decisions, and long-term intent. They understand that transformation is not measured by platforms deployed, but by clarity gained. Not by velocity alone, but by coherence sustained over time.
If there is a destination to Digital Transformation, it is not technological mastery. It is human empowerment. It is the quiet confidence of teams who understand their tools, trust their leaders, and act with autonomy. It is partnerships strengthened by transparency rather than control. It is value chains that adapt without panic.

This book does not conclude with an answer, because leadership rarely does. It concludes with a responsibility: to see clearly, to act deliberately, and to remember that in a world of accelerating change, the most enduring advantage remains thoughtful leadership.

'

About the Author

Satya Sanivarapu is a technology and business leader whose work sits at the intersection of enterprise systems, digital transformation, and organizational clarity. Over nearly two decades, he has been working across Supply Chains of Global Manufacturing, Retail, Logistics, and Consumer Goods landscape, guiding complex Enterprise Resource Planning, Warehouse Management, and Analytics ecosystems through periods of modernization and change. His experience spans both industry leadership roles and advisory work through Sansid Consulting, where he focuses on helping organizations place digital capability thoughtfully—balancing technological possibility with human judgment. Educated in Engineering, Business, and Supply Chain Management, Satya's perspective is shaped by long exposure to large-scale systems and the quiet realities of leadership. He lives and works in New Jersey.